CW01463903

VENEERING
AND INLAYING

A Study of Materials,
Principles and Processes

By

G. M. Nyman

Published by
The Woodward High School Printing Class
Cincinnati

Veneering and Inlaying

The art of enriching objects by the use of rare and costly woods in the form of veneering was known and skilfully employed by artisans of a remote age.

In European museums and, undoubtedly also among collections of antique art in this country, curious examples of such work can be found. In many instances the veneer is fastened to the core with small wooden pegs and from this we draw the conclusion that glue was not known at that time, at least, not as a medium for fastening veneer to the core. The early craftsman method of veneering, judging from their work, must have been somewhat as follows:

After having cut his veneer to a thickness of from $\frac{1}{8}$" to $\frac{1}{4}$", he held it down on the object to be decorated and bored or cut small holes thru the veneer and down into the core a short distance. In these holes he inserted small pins or pegs made of the same material as the veneer. These pegs were wedge-shaped, with an elliptical head and concave sides coming to a point. When driven home, the points were firmly imbedded in the core, thus holding the veneer in place. Corners and edges are often reinforced with metal, this preventing the veneer from being stripped off, as well as serving a decorative purpose.

Later on, when the use of glue became known, veneering made rapid progress and some of the best pieces of antique handed down to us bear mute evidence of artistic skill and patience in the employment of veneering and inlaying of wood, metal, and other materials for surface decoration.

Some of the furniture known to us as "Period" furniture could not have been executed had not the old masters known the art of "building up"—gluing three or more layers of wood together in such a way that the grain in one layer always runs at right angles to that of the adjacent layer.

The employment of veneers forms a very important part of the woodworking industries today and some knowledge of the processes and principles used in this work will benefit all who work in wood and use glue.

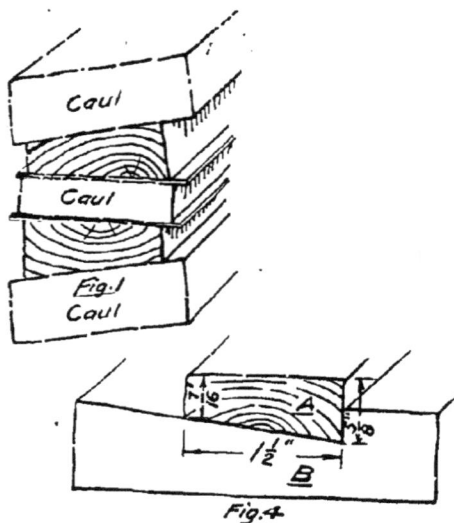

Fig. 4 shows how a simple and pretty form of

small picture molding can be made where machines are available, or by hand.

The shoe *B* is rabbeted on the circular saw and the desired angle obtained. The mouldings are ripped to width, then placed on the shoe and run thru planer. This will give moulding the proper shape. The best and most economical way of making them is to veneer wide boards, which can then be ripped up into mouldings of any desired width. Mouldings more than six feet long are cumbersome to handle and should be cut.

One thickness of veneer is all that is needed for these little mouldings and that is cut in strips across the grain ¼" wider than the width of mouldings. In cutting, lay sheet of veneer on a board, mark off strips and cut after a straight edge with sharp knife or chisel.

If veneer is brittle and has a tendency to split, dip a brush in water and dampen the veneer slightly. Lay out the strips in the order of cutting, join ends and match.

In matching the aim is either to produce a strip

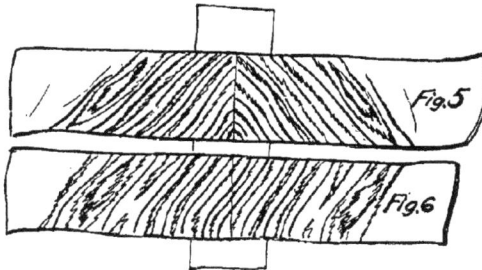

of veneer seemingly without joints as in Fig. 6, or by turning one of the members over, a matched effect is obtained as in Fig. 5. The joints are held together

3

with paper tape. Long joints are nailed down to a board with small brads previous to taping, but short joints can be held down with the fingers while the paper tape is glued on. When tape is dry, the veneer is ready to lay.

Fig. 1 shows arrangements of mouldings and cauls. The center caul is greased and heated while glue is applied to the mouldings. The veneers are then stretched over the glued mouldings with paper-taped joints on the outside. At this stage the center caul is taken from its heating place and put between mouldings to be veneered—the veneered sides facing the warmed caul. The outside cauls are then placed and pressure applied. These outside cauls could be eliminated on larger mouldings of same type, but are used in this instance to assist in even distribution of the pressure, and to prevent clamp markings on back of mouldings.

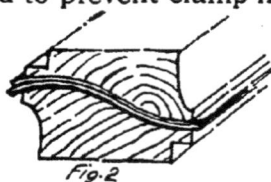

Fig. 2

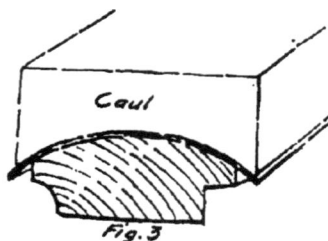

Caul

Fig. 3

Fig. 2 suggests a moulding without the use of center caul. In this and similar cases, the mouldings themselves are heated and the glue applied to them. The warm moulding

4

will absorb considerable glue and this must be taken into consideration when glue is spread. Glue can also be spread on face veneer but this should be avoided if possible. (See chapter on glue.)

Fig. 3 is a suggestion for a plain flat-round moulding that will look well when veneered. It is advisable to veneer all these little mouldings before glass rab - bet is cut, as there is danger of breaking same in the veneering if made previous to that.

To make these and similiar mouldings by hand run some 1¼" screws thru a 1"board, place the screws in line about a foot apart. Then file the protruding screw points so they will grip the blank mouldings when hammered down on them. The mouldings are then planned to the desired shape with a moulding plane or in case of the round moulding illustrated, an old wooden smoothing plane can be converted into a moulding plane and will do good service.

Built up panels are bought ready made and used for different purposes in many school shops, and rightly so. However, the value, educationally and otherwise, would be far greater if those who use built up panels also had a chance to make them and there are few schools so poorly equipped that they cannot turn out some work of this kind. Take serving tray bottoms for example. All that is needed are two pieces of 2" surface planking for cauls of the length and width of tray contemplated (rarely more than 15 by20); two face veneers (one for top and one for bottom)and a 31-6" poplar or birch core. The grain of the latter runs at right angles with the face veneers. Heat cauls a little while glue is spread on both sides of core, which is then placed between face veneers

5

and the whole placed between the greased and heated cauls, and pressure applied with heavy hand screws or clamps. . If the cauls should have warped some during the heating, place the convex sides to the veener, thus insuring pressure in the center first. About a dozen hand screws or clamps will give sufficient pressure for a panel of the size mentioned above. After six hours the hand screws can be removed and tray bottom taken out, trimmed to size and inlaid if desired.

To cut grooves for borders in veneered work, make a tool like the one sketched in Fig. 7, the

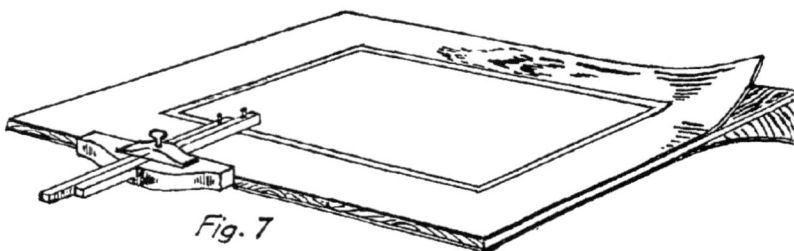

Fig. 7

most important parts of which are the cutters. These should be knife pointed so as to cut equally well across or with the grain. An old gauge block with set-screw arranged as in Fig. 7 will take care of the adjustments.

When face veneer is cut thru, the strips are removed with a small chisel, borders or lines fitted in and glued under pressure.

It takes three days for a veneered panel to dry and it should be kept clamped between sticks during this time, so that it will not warp while drying.

Face veneer is knife cut and smooth to start with and it does not take much scraping or sand papering to make it ready for finishing.

6

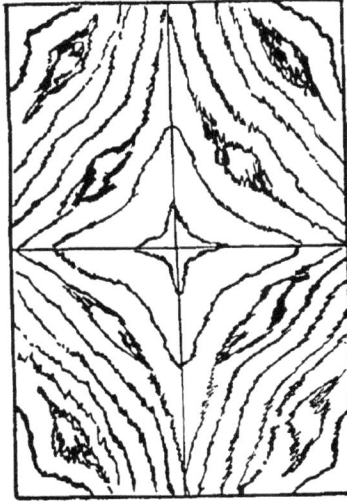

Fig. X

Matched panels, as illustrated in Fig. X, should be made five-ply in order to stand up. The gluing of a five-ply panel is no more difficult than the three-ply previously described. In order to make such a panel it is necessary to have four successive cuts of veneer from the same log in order to match the face veneers. The veneers are then joined, fastened on the boards with brads and the joints taped as previously described.

Checker boards are very popular with school boys, but if they are glued up of solid blocks without any backing, the life of the board will be short, as the expansion and contraction will open up the joints. Five-ply veneer checker boards have proven to be more permanent and are not hard to make.

Two pieces of veneer, one dark and one light, are ripped into eight strips, each 1½"x14". It is important that these strips are of the same width and in order to get uniform widths, make a jig for that purpose as illustrated in Fig. Y. If a fine power saw is available, the strips can be ripped on same.

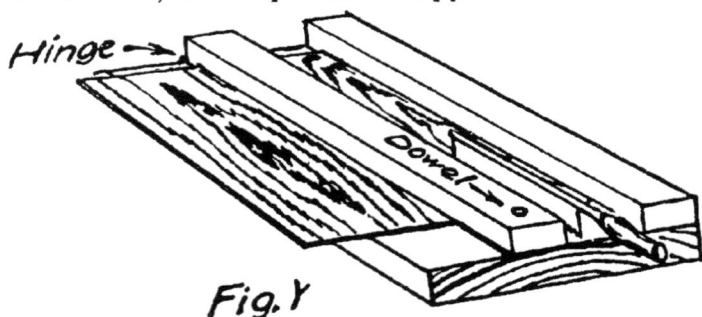

Fig. Y

The strips are then alternated, fastened to a board with brads and paper tape glued over the joints. When dry cut off eight strips across the grain

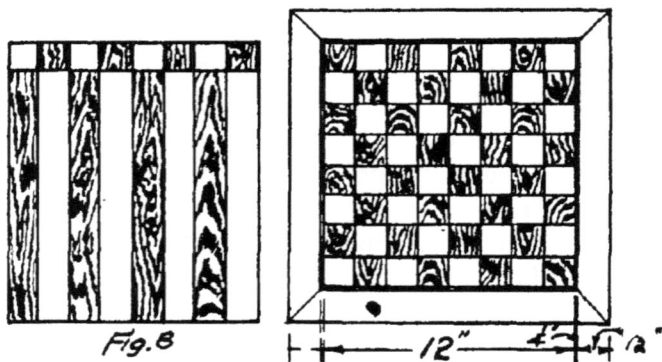

Fig. 8

12″

(see top of Fig. 8), alternate and tape together again as shown. When dry, join edges and mitre borders or lines around, fastening same with paper tape to the checkers. This finishes the face veneer of checker board. Now cut back veneer, two cross bandings and core of same size as face veneer, and glue together

8

as shown in Fig. 9.

Table rails are easy to veneer and will serve as well as solid ones if they are made like Fig. 10. They can also be veneered with a single veneer, but they

Fig.10   Fig.9

have a tendency to warp. Table legs glued up and

Fig. Z

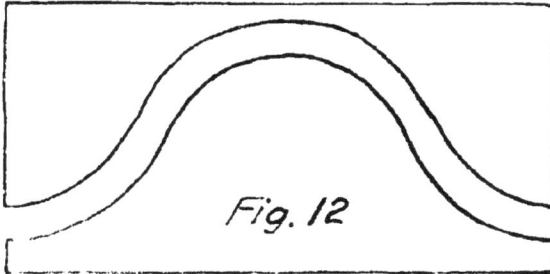

Fig. 12

9

veneered with ⅛" veneer (see Fig Z) have many
points in their favor.

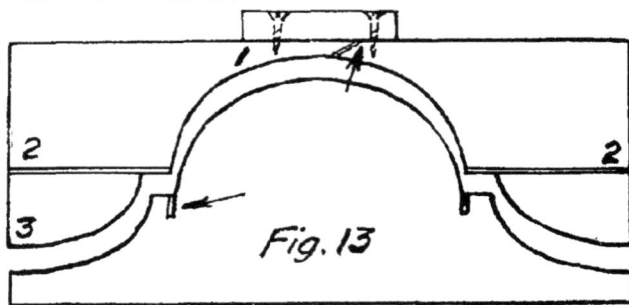
Fig. 13

The veneering of curved objects is more difficult.
Figs. 12 and 13 show some mounts for small clocks.
These must be band-sawed carefully so that the waste
pieces can be used for cauls.   Fig. 12 is the least
difficult to veneer and the curved can be done in
one operation.   In this case, it is best to veneer ends

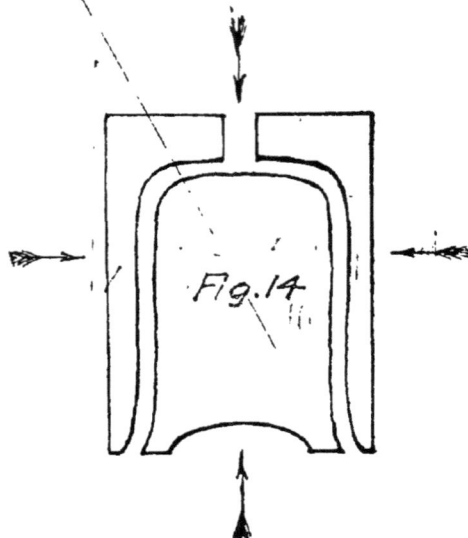
Fig. 14

first, then curved part and finally the face. Fig. 13 is more difficult to saw out and to veneer. The first cuts are made at Fig. 2, next cut from arrowhead on top to other arrowhead, then at Fig. 3. The veneering is done in this order: The veneer is let into kerf cut at lower arrowhead and glued in place. The ends, covets, fillets and face are veneered in the order mentioned.

In order to veneer pieces having curves of small radius, (Figs. 13 and 14) it is necessary to dampen and bend the veneer to the approximate shape of the object. This is readily done by dipping the veneer in water and bending it over a hot steam or gas pipe. If there is a little glue in the water, the veneer will keep its shape better.

Arrowheads in Fig. 14 indicate the direction of

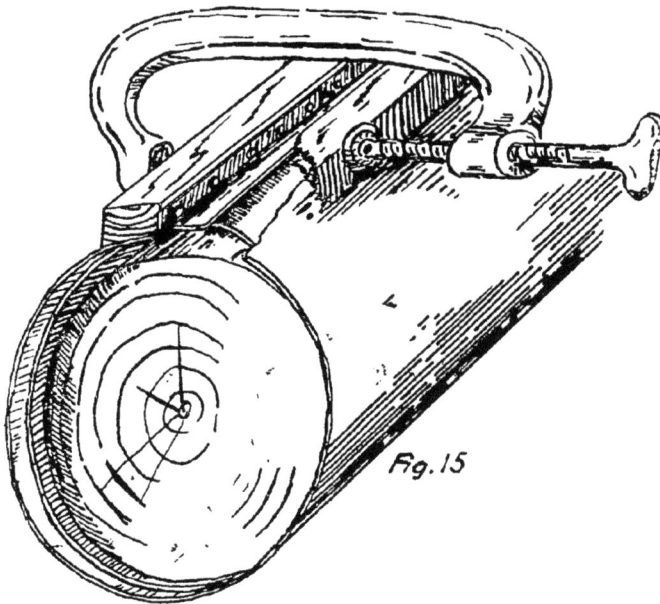

Fig. 15

pressure to be applied when veneering.

No dimensions are given for these clock suggestions because it will be found best to get the movements first and make the cases to fit them. Any jeweler can get movements from makers or wholesalers and these range in price from 75 cents up.

Fig. 15 illustrates the method of veneering round columns. A piece of sheet metal, alumined or zinc with two strips of wood securely fastened to the edges is heated and clamped around the object to be veneered, leaving a little space where veneer laps without pressure. This later reheated, jointed and pressed down.

Veneers range in price from a fraction of a cent to 80 cents per foot or more. Dealers are located in all large cities and they accumulate what are known as "dead samples," meaning only a few of the same log on hand. These can be purchased for one to two cents a foot. Cross banding costs from one-quarter cent up. Poplar core costs about one-half cent a foot.

The Ohio Veneer Co., of Cincinnati, O., will be glad to supply veneers. J. B. Bernard Co., 422-30 East 53rd St.. New York City, will be pleased to supply lines, borders and insets at a reasonable price.

Minimum of Equipment.

No elaborate equipment is necessary and good results have been obtained on small articles, such as tray bottoms, fancy boxes, checker boards and picture frames in school shops where clamps were the only means of pressure, also using cold glue. In our shop we have successfuly veneered stock up to 24" by 36", using ordinary handscrews, hot glue and heating our cauls on the steam radiator.

Constituents of Veneered Work.

Veneered work is usually found to consist of three or more layers, core, cross banding and face veneer. The core stock may be 3-16" or any thickness suitable to the purpose. If thin, it is usually rotary-cut poplar or birch. If thick, any of the following woods are likely to be used in order mentioned: Chestnut, poplar, birch, soft pine, elm, or others prevalent in the locality.

Core Stock.

Good gluing and staying qualities are desirable in the core stock. All defects in same should be remedied – large holes plugged and smaller ones filled with a paste made of silex and thin glue (charcoal, chalk and thin glue also makes a good filler). The filler must be dry and is then tooth planed level with surface before veneering.

Cross Banding.

The cross banding (1-16" rotary cut poplar or birch) serves a double purpose—that of stiffening the object, preventing it from warping and shrinking and also forms a soft, even-grained surface for adherence of the veneer.

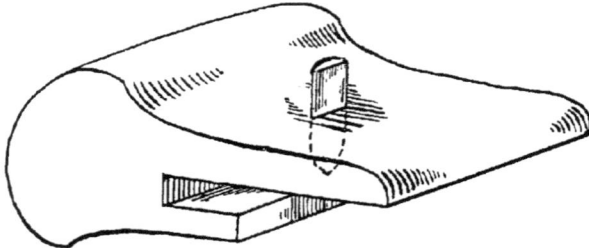

Advantage of Good Blind Stock.

On highly polished furniture. where the face veneer is laid on coarser wood, one can often see every line in the grain of the underlying wood—the softer spring growth will shrink more than the later growth knots, being harder than the surrounding wood, will stand out especially prominent. The cross banding

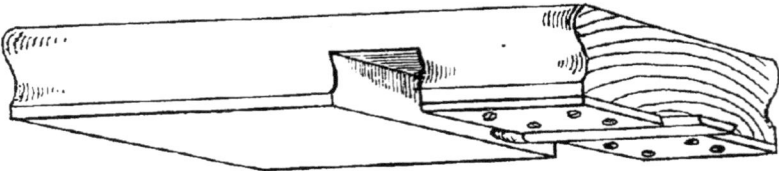

will act as a medium preventing this, altho if veneering work is not given time to dry thoroly before being cleaned up and finished, it is likely to show it later on in spite of cross banding. Many manufacturers glue a coarse cheesecloth under the face veneer, thus preventing the grain of the core wood showing thru

14

and also giving better gluing surface. However, this is only done on fine work.

Joining and Taping of Veneers

In case the crossbanding or face veneer has to be joined, place veneers between two boards, held in place with a handscrew at each end, and fasten in

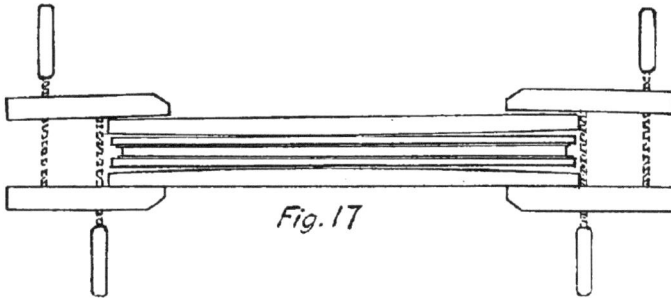

Fig. 17

bench vise for joining. Good veneer joints can be made with little effort by the use of a shooting board and a sharp jointer plane. The veneer is held flat on the board with a stick and joined. The pieces are then fastened on a board with small brads (½ No. 20) so as to keep the joints tight and even while tape is glued on. For cross banding and thin core

Fig. 16

use cheesecloth tape; for face veneer, paper tape. The cloth tape on the cross banding is merely to prevent the joints from slipping apart or lapping, and a short strip glued across the joint at each end will

15

do the work. However, the leaving of a few patches of thin tape at intervals under the face veneer, if they are needed, does not affect the quality of the work.

On face veneers tape all joints the entire length, also tape all split places to prevent futher rupture. In many cases ends as well are taped to insure safe handling. Cheesecloth and paper tape are on the market. The latter can be secured from any paper dealer, but the home-made article will serve all purposes. Use thin, tough paper cut in strips three-fourths to one inch wide.

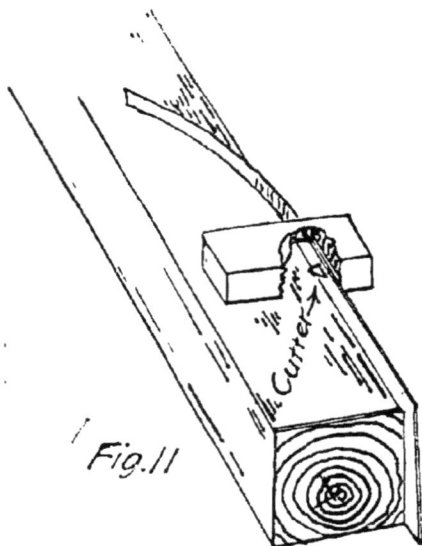

Fig. 11

Remedying Defects in Face Veneer

On fancy veneers (walnut root and crotch mahogany) there are often found small defects—little chips missing or tiny holes. These can be filled with a paste made of scrapings of the same wood and very thin glue. Coloring matter can be added to give the filler any desired shade. Yellow ocher will be found to work well where a lighter shade is desired. This paste kept in a wide necked bottle and applied on the little defects before the veneer is laid will save subsequent labor. If the holes are of any size, pieces of veneer to match should be fitted in and bits of paper glued over.

Cauls—Metal and Wood

Among the important factors in veneering are the cauls——wooden boards or metal sheets as the case may be— used for reheating the congealed glue and keeping the same in a liquid state until pressure has been applied. Metal cauls (aluminum or zinc) are by all means the best and most economical. They should be thick enough to serve the purpose stated above, so that if it takes a long time to get the job under pressure, the metal will be thick enough to retain its heat for that length of time. One-sixteenth

17

inch (1-16") thickness is satisfactory for all-around flat work; curved work requires more flexible metal—24-gauge. The cauls in this case are backed up with wood and both metal and wood are heated. This is also done on flat work, when the metal is too thin to retain heat for the required length of time. The sooner the caul cools, after pressure is applied, the better. A new batch can then be put into the press. Overheated cauls are detrimental to good work, as they cause undue expansion of the veneer and the absorption of glue into the wood, which, at times, leaves a starved joint. Keep cauls greased with tallow to prevent them from sticking to the veneer. Wooden cauls are satisfactory when made of material

that will stand repeated heating without warping. Chestnut, butternut and pine are such woods. In addition to greasing wooden cauls ,it is well to place newspaper between them and the veneer.

The cauls should be of uniform thickness—about ¾ inch. This will allow two cauls to be placed side by side in case a larger surface is to be veneered. If there are facilities for heating the cauls, as many pieces can be veneered as the press or handscrews will hold. If the radiator has to be depended upon for heating of cauls, it is not advisable to veneer more than one piece at a time. Composition boards have been tried for cauls, but are not finding much favor, but three-ply panels about ⅜ inch in thickness are

used in many places.

Where band-sawed work is veneered, the waste pieces are used as cauls. This, of course, necessitates close band-sawing and, should th is not be entirely successful, the bumps must be taken out of the piece to be veneered, as well asthe waste part, and a thickness of felt placed between to equalize the pressure. The usual precautions should be taken to prevent the felt sticking to the veneer. Use a thin piece of zinc or paper. Such problems as the edges of round tables are invariably veneered by having a metal band drawn tightly around. This band is a little shorter than the circumfreence of object to allow for clamping. When everything is ready and the veneer

is in place, the band (now warm) is drawn up tightly. That part between blocks not pressed down, is later on softened up with water and a hot flat iron, and then rubbed down with a veneer hammer.

Table rails, or other one-piece circular objects glued up of segments, are not hard to veneer if the work is gone about in the right way. In many cases the segments glued up are wide enough to allow ½ inch to ¾ inch to be cut away when trimmed to size. This waste part is used for caul. A board, thin enough to bend to the desired curve without breaking, can also be used for the same purpose. In either case the cauls must be heated and handscrews applied 4 inches to 6 inches apart, beginning at one point, going around and ending up at starting point.

Glue.

Glue used for veneering should be thicker than that used for ordinary jointing. The proportion of glue to water depends largely on the quality of glue— the better grades of glue will stand more water and give more glue of a given consistency, pound for pound, than the cheaper grades, thus proving really more economical, besides giving better all-round satisfaction.

A veneer joint is not different from any other glue joint, except that the glue is spread over a larger surface and the veneer, being thin and easily affected by heat and moisture accompanying the gluing process, requires careful handling.

The consistency of the glue used in jointing should, to a large extent, depend upon the absorbing qualities and texture of the wood used. It should be thin enough to flow readily into all cells and ducts in the surface of members forming the joint, yet it

21

should be of such a consistency that, when dry, it leaves sufficient solid glue matter to connect the cells of one member with those of the other, thus dovetailing them together and forming a union that cannot be broken without rupturing the cell walls.

Glue size is glue thinned with water until the presence of glue is hardly perceptible, when run from brush, between fingers. It is used preparatory to the veneering of curved work having more or less end grain and on very soft core stock, where the veneer glue is likely to be absorbed, thus leaving the joint without the proper binding substance.

Things to Consider in Order to Obtain Good Results.

Good results in veneering can be obtained by following these suggestions: Use glue of proper consistency; have the cauls just warm enough to do the work intended and get the batch under pressure with the least expansion of veneers.

Good veneer work should be free from blisters and lumps—should have no minute checks, nor should it warp. The defects mentioned are hard to overcome after once created, but are easily prevented. Blisters may be caused by too thin glue, or not enough of glue; by overheated cauls, or by a combination of all. Unevenly distributed pressure may cause blisters, so will caul that is not warm enough to do the work. (See cauls.) The latter may also cause the surplus glue, not hot enough to be pressed out, to collect in some places under the veneer, thus forming high spots on the face veneer which, when leveled, will leave the veneer very thin— even running the danger of smoothing thru.

Hair-line checks are usually caused by undue expansion of the face veneer. This veneer, glued

22

down and made part of the whole, has no chance to regain its former dimensions but the strain produced will cause the minute checks to occur. They are, in most cases, not visible before the piece has been finished and the damage done. The more expensive veneers are most likely to check—such as crotch mahogany, in which part of the grain runs almost perpendicular to the surface. This end grain will quickly absorb glue and expand, the strain in subsequent contraction causing little checks.

To prevent undue expansion, used rather thick glue and do not lay the veneer on the hot glue, but wait until the glue congeals. If necessary help this chilling action with a fan. Have cauls moderately hot and do not veneer too many pieces at a time, thus avoiding delay in getting the batch under pressure.

Veneers are usually cut to extend about ½" outside the piece to be veneered. This is done to take care of possible sliding. If the veneer should be scant, a small brad fastened in the middle of each end and bent over will prevent sliding. However if the pressure is put on gradually, first in the middle to press out any surplus glue and then all around, there is little danger of sliding. It will be necessary to tighten the same screws more than once to insure even pressure.

How Veneer Is Affected by Glue

As hot glue is spread on thin veneer, the side receiving the glue will immediately begin to expand. This action is so rapid that in ten minutes the flat veneer may have formed a complete tube, due to one-sided expansion of the veneer. If let alone to dry, the tube will gradually flatten out and then begin to curl in the opposite direction and if left over night, the contraction of the glue will cause the tube to turn inside out.

23

It is evident, therefore, that the glue should be spread onto the thickest member of the veneer joint (the core). However, if cross banding is used, it is customary to spread it on both sides. The first side is laid down on narrow strips (triangular shape prefereble) to avoid rubbing off glue on this side while the other side is being spread. One side of the cross banding is laid on the core; the other side receiving the face veneer. Each side of the cross banding must have sufficient glue to cover the adjacent unglued surface, with some to spare. The glue should be spread on—not painted, as beginners have a tendency to do. The latter process may cause starved joints with resultant blisters.

Hints for Prevention of Warping and Checking of Veneered Stock

The thicker grades of sliced and rotary-cut veneer are more or less ruptured as the log revolves against the knife and it is important to lay such veneer with the ruptured side down; in this way avoiding small checks in the finished article.

The warping of veneered stock may be traced to a number of causes. One or the other of those mentioned below may be responsible or perhaps a combination of them:

The veneer on the back not being equal in strength and pulling power to the one on face.

The caul on one side may have been warmer and caused more expansion than the one on the other side.

The grain in one veneer running diagonally may give a twist to a thin panel.

In most cases, however, the warping is caused by the core being the strongest single member; it may warp the whole according to its tendencies.

If, as in thin panels, the core is rotary-cut ⅛" poplar or birch, the tendency is to curl in the way it was cut from the log, but in many cases this can be used to advantage. Cores, built up of strips (laminated) will not warp, nor will quarter-sawed woods. Plain sawed wood will incorporate the same tendencies in a veneered job as when used by itself.

Application of Glue and Pressure

When about to veneer have sufficient glue on hand to do the work intended. To run short is an exasperating situation. Also have enough cauls for the job—this means one more than the number of pieces to be veneered. The uppermost one must be thick enough (See Cauls) to distribute the pressure. If handscrews are used, the bottom caul must be thick also, for the same reason. Cauls, as a rule, are kept in uniform thickness. Added stiffness can be obtained by placing an extra caul on top, or if metal cauls are used, the last one is backed up with

wood. Have all cauls warmed and greased. Esti-
mate the thickness of the batch to be veneered and
adjust screws to the approximate size. The pressure
is applied to cross pieces—common two by fours—
put on top of the whole; on bottom also if handscrews
are used.

In case wide pieces are to be veneered by the use
of handscrews, some precaution must be taken to
insure pressure in the middle, where the handscrews
will not reach. This is done by rounding the cross
pieces in such a way that they will press on the center
first. About one-quarter inch at the ends rounded
to nothing will do the work. Alternate rounded
and flat cross pieces, two of a kind opposing each
other. This will insure even distribution of the

GLUE PRESS IN USE

The sections and bottom are made of 3 in. beech which is unfit for
ordinary use on account of checks, but strong enough for such a press.
The cut shows the press in use without the screws, the pressure being obtained by blocks and wedges, but as soon as the boys in the machine shop
have finished the nuts and screws, they will be hung 3 to each section.

pressure.

The Care of Stock---Veneered and Otherwise

It takes about three days for a batch of veneered work to dry. During this time, it should be piled on sticks, or the top piece covered to insure even drying. Stock, veneered on one side, should have that side covered, as slow drying will retard possible warping and checking.

In this connection a few words, regarding the

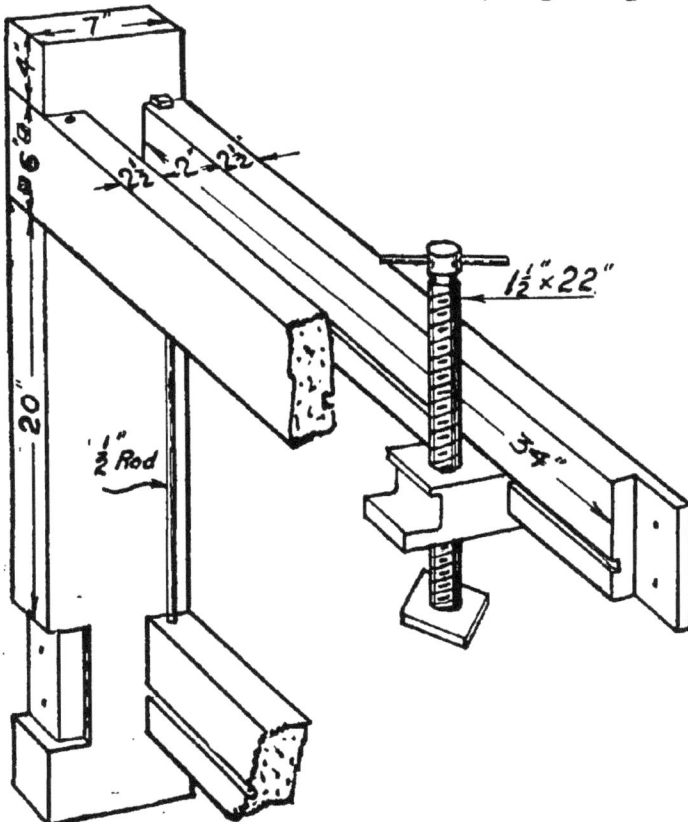

Sketch for one section of Veneer Glue Press.

general care of work under construction in school shops, will perhaps be helpful. As a rule the atmosphere in the shop is dryer than the lumber used and until the moisture in the shop and lumber is equalized, the seasoning of the latter will go on. As the moisture evaporates, the lumber will shrink. A board, evenly exposed to air on both sides, should, according to theory, remain straight and it will in most cases if the lumber is quarter-sawed. Plain lumber, however, will warp under the same conditions, causing the bark side to be more or less concave, with a corresponding bulge on the heart or pith side. Hence all such lumber should be put away with the pith sides exposed to the drying influence of the air. The pull, then caused by the shrinkage of those sides, is

Veneered Cabinet made in the Author's Classes.

opposed by the strong resistance of the bark sides and the lumber usually remains straight.

The Rubbing Method.

Another method of veneering—second in importance to the one previously mentioned—is the rubbing method, getting its name from the way the veneer is rubbed down and surplus glue rubbed out in a series of strokes with the broad pien of a veneer hammer.

This process requires more skill than the pressure method and is done in the following manner:

The core is tooth planed and glue sized. When dry, it is again tooth planed slightly to remove any little bumps. In the meantime the glue is prepared; two flat-irons heated; face veneer cut and dampened down and the glue is then applied to the entire surface of the core—if same is not too large. Larger surfaces are spread and rubbed down in sections. The face veneer is then put in place and one of the flat-irons applied to face veneer in middle of one end. The hot iron will reheat the glue on as large a surface as is convenient to rub down with the veneer hammer before the glue gets chilled. Proceed with the rubbing in this way:

Hold the veneer in place with one hand while hammer is rubbed on warmed spot with other hand. The first strokes are generally made from eight inches to ten inches in toward the end. This should bring out some surplus glue, indicating that the conditions are right. If not, moisten veneer on top and apply flat-iron until the desired result is obtained—removal of surplus glue and air. The veneer then will adhere to the core. As soon as this is accomplished in one place, the veneer hammer is grasped in both hands—

29

the right pulling and steadying the handle, while the left is pressed down on the hammer head. The motions are with the grain and diagonal—all the while pressing the veneer down and the glue forward and to the sides. Apply the flat-iron and reheat another section, repeat and the job will be finished in a short time.

Apply no more moisture than is needed and avoid stretching the veneer as it is rubbed down. The greatest drawback to this process is the expansion, with accompanying evils. While this method of veneering is mostly relegated to the fixing of imperfect glue-press work and special complex curve work, it is still used and therefore mentioned here.

From the school shop standpoint, it has the pleasing feature of not requiring many additional tools. In fact, the only tools it will be necessary to acquire are two flat-irons and a veneer hammer. The remainder—chisel, square, T--bevel, and a straight edge—are found in most shops. Supplies needed are hot glue and veneer.

Knife–cut mahogany is easiest to lay by this method; however, a skilled operator can lay any kind of thin veneer.

In our shop the boys have successfully veneered several large floor lamps, employing the rubbing method. The lamps, in question were 5 feet high, had a slender tapering body, increasing in size to twelve inches at the base. Being too large for accurate band-sawing, the lamps were planed to proper curve and veneered by the foregoing method.

Veneering of Frames

Frames are readily veneered and corners mitered in this manner: Veneer the stiles first, then the rails —the veneer on the ends of latter overlapping the former.   Enough of the surplus veneer is then trimmed off on the inside of the frame to allow a rest for 45-degree angle, or T-bevel as the case may be, and the overlapping veneers are cut thru with a thin sharp chisel. The waste part of the topmost veneer is removed without effort, but in order to re-

move waste part of underlying veneer, it is necessary to reheat this spot with the flat-iron, turn up the end and remove underneath waste piece, putting a little glue in its place. The turned-up end is now rubbed in place and the joint is taped to prevent opening while drying . A mitre of this kind is bound to fit as the two members are cut at the same time.

Friezes and borders of panels; drawers, etc. are done in the same manner.

One advantage of this method is that the work is visible at all times.

How to Make Concave and Convex Picture Mouldings.

The contracting pull of a one-side veneered surface is something that everyone doing this work and giving it some thought cannot help becoming familiar with.

It is sometimes taken advantage of in making mouldings when a cove is desired. The process is as follows:

A thin strip—¼" of some soft, pliable wood— is veneered. The veneer previously has been expanded by the application of water. In drying the

32

veneer will contract and in doing so will turn the core from a flat piece into a hollow cove. This, as previously mentioned, is nothing new, but to turn a flat picture moulding of good sized dimensions into ʋ handsome convex curved one is a novelty that it was our good fortune to discover.

The moulding in question, intended to frame a 28"x40" picture was 6"wide, 1¼"thick at outer edge, tapering to ⅝" at inner. A poplar plank 2" thick was the stock used. This, with others, had recently arrived from the lumber yard and was dry, from the yard people's standpoint, which means dry enough to work but containing considerable moisture.

This plank was surfaced on four sides and then allowed to stand in the shop for a week while the veneer was prepared. As the steam heat was on, the surface of the plank dried at that time. It was then re-sawed at an angle, giving two mouldings of the desired shape. The re-sawed surfaces were planed and veneered on the sides that had been allowed to

33

dry, with striped mahogany, the grain running at right angles to the core. The next day all hand screws were taken off, the surplus glue removed and the two mouldings put away to dry, held by a C clamp in each end, veneered faces together and the re-sawed surfaces exposed to dry, with resultant shrinking and warping of those surfaces. In about a month, the mouldings had aquired their permanent shape, which on the face showed a prettry convex curve, being more pronounced at the thinner part of the tapered moulding.

Inlaying of Lines and Borders.

The inlaying of lines and borders on veneered work is comparatively easy. A tool, based on the principle of the double-pointed marking gauge, will do the work. The block should have a long face resting on the edge of object to be cut. This will insure a straight cut, if held firmly against the edge as it is moved along. The cutting points are the most important. They should be made to cut across the grain, as well as with the grain, and knife points will be found to do the best work. After the veneer is cut thru, the strip is easily removed. Especially is this true if the cuts are made before the face veneer is completely dry. In the latter case a narrow chisel can be inserted under the cut veneer and the strip will come out in its entire length and the line or border glued in its place—either pressed down or rubbed in place with a veneer hammer.

Small centerpieces, such as sunbursts, sea-shells and wreaths are sold inserted in squares or rectangular pieces of veneer. These can be trimmed to any desired shape, after which they are laid in position and scribed around; the background is then removed and the insert glued in place under pressure.

Furniture factories, using large quantities of lines—plain black and white—usually cut their own

Veneered Music Cabinet.

from logs of ebony and holly, using fine circular saws. Fancy borders of rare colored woods—natural and dyed—are purchased from manufacturers making a specialty of this kind of work. The borders come in yard lengths and retail for four cents a yard and up. The same manufacturers will also furnish lines from 1-16" to ¼" in black and white woods, natural or dyed, as well as celluloid, brass and copper lines.

The most popular single border consists of white lines between two darker ones or vice versa. This can readily be made by gluing three veneers, having

the desired colors, together and sawing it in strips. In this way a number of more or less intricate designs in borders can be made up by those having the time, inclination and material. The latter can easily be purchased.

Marquetry

The art of reproducing flowers, fruits and other forms of nature, as well as objects of human design, is called marquetry. The materials used are colored woods of varied kinds in there natural state or dyed, besides shells and thin metals.

The marquetry cutter must be a person of great skill and artistic ability in order to work out the

36

pictures true to life in such unresponsive materials. The work is done by tracing the design on the veneer which is to form the background. The other veneers to form the design are glued onto the back of the former, with paper between. They are then cut with a marquetry cutter's saw—a machine resembling a foot-power scroll saw, adjusted with a tilting table or a tilting saw. This is to take care of the space left by the saw cutting its way thru the materials. As very fine saw blades are used, it does not take

Veneered Marquetry Panel.

much of a tilt to close kerf when design slips into place.

After being sawed the design is removed and shaded. This is done by dipping the piece in hot sand and scorching until the desired depth is obtained. Afterward the background and all pieces forming the design are assembled and glued onto a soft piece of cardboard and the marquetry is now ready for the market.

Amateurs can do a little experimental work along this line by embellishing objects with simple conventional design. A fine scroll-saw can be used; circular parts can be punched, or bored out with a Forstner bit. Lay all marquetry with paper side up. When dry after two days, remove papers by water soaking and scraping.

Marquetry work—borders and lines—rightly applied add distinction and beauty to any object. It sets off and give life to surfaces that, without any embellishment, would appear flat and monotonous.

37

As the work itself requires considerable skill in execution and application, the presence of it stamps the article as valuable.

Displaced and overloaded application of this kind of work will cheapen the appearance of any object, as will inharmonious and loud effects in coloring and outline.

This method was formerly used for work having complex curves, but is now obsolete.

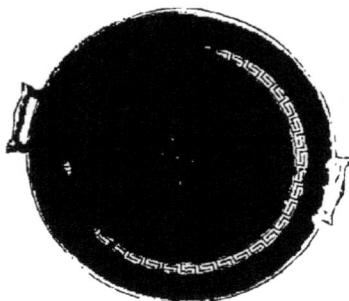

Veneered Tray

To veneer by this process, fill with sand a box of proper size for the work to be done. Press piece into the sand until an impression is made. The sand is then slightly heated, after which glue and veneer are applied and the whole pressed into the imprint previously made in sand box, and clamped down.

Methods of Cutting Veneer

A variation in thickness is often found in fancy veneers when they come from the manufacturers. In order to insure even pressure when laying, it is necessary to build up the thin veneers with paper or thin cardboard until uniformity is reached.

38

Inlayed Pieces made in the Woodward High School.

Face veneers for cabinet work in the market are cut from 28 to 30 sheets to an inch, the thicker grades to be preferred as they will allow for smoothing off. As many sa 160 sheets to an inch are sliced from Spanish cedar. This thin stock is used for the veneering of cigar boxes.

The slicing method of cutting veneers is perhaps the most common. The steamed square log is clamped in position, the knife comes down vertically and cuts off a slice with a shearing cut, after which the carriage moves the log in position for a new cut and so on until the entire log is disposed of, with the exception of that part held by steel dogs to the carriage.

Rotary cut veneer, so called when round log is held between centers, revolves against the stationary knife and a continuous sheet is turned off. Another method is when the round log is cut in two and the halves, one at a time, revolve against the knife, producing a semi-circular cut on the round side of the halved log.

The above methods require boiling of the logs before cutting. This, however, is not done when the logs are sawed into veneers, as most of the thicker grades are.

Fancy root veneers are sometimes so undulated that they cannot be matched and joined, nor laid without first having been dampened down and pressed between hot wooden cauls. This will take out most of the kinks and make the veneer easy to work.

Specially constructed circular saws of large diameter are used in veneer sawing. These saws waste very little lumber in cutting. They are made with a

40

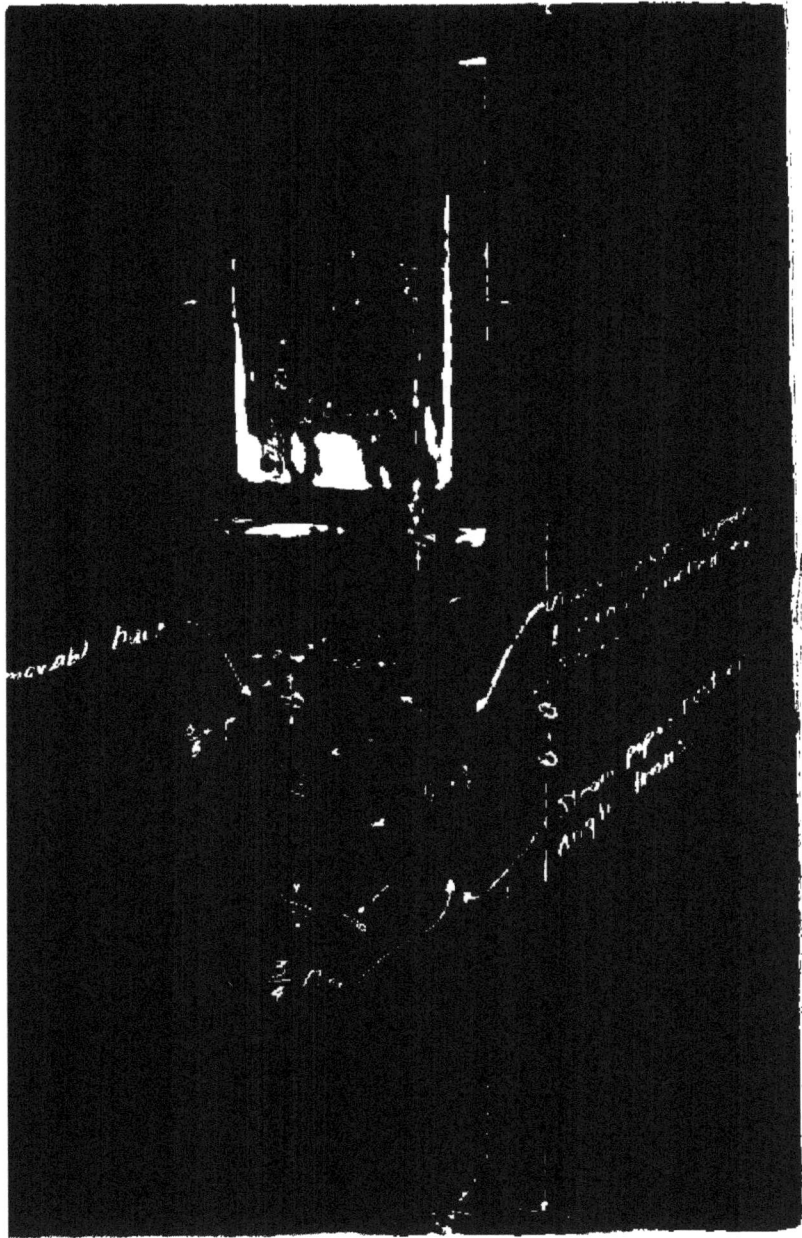

large iron disk, which is quite thick at the center; the cutting rim is fastened to this in sections with machine screws. The rim, itself, is hollow ground to 1-32" at the cutting point—1-32" therefore represents the amount of waste for each cut. It is evident that the saw must be perpendicular on the side towards the log; the increasing thickness from teeth to arbor all occurs on the outside. As many as twenty sheets to an inch are cut by this method.

Veneering has made it possible for us to enjoy beautiful furniture; it has a place of its own and needs no apology for its existence.

People are beginning to realize that veneered articles do not necessarily mean cheap articles. In fact, a built-up panel will, in most cases, cost more than the solid one. Matched panels of fancy figure would be entirely out of the question in solid wood; the changing structure of the log for each succeeding cut would forbid that, even if it were possible to build furniture of wood with the grain running in all directions.

Preparing Veneered Surfaces for Sanding and Finishing.

In two or three days after the veneer has been laid, the piece or pieces are taken out of the press and all paper tape and glue are removed and the surface cleaned off. This is accomplished with sponge, hot water and scrapers suitable for the work on hand. This work of cleaning deserves more attention than it receives in many places because the hot water will not only remove paper and possible surface glue, but will shorten the time of subsequent sanding and lengthen the life of the sandpaper. The warm water will show up blisters, if any, by causing a bulge on

42

the face veneer wherever that does not firmly adhere. Of course, this is not of much importance in the school shop; in the factory, however, where production is the aim, it is worth looking after.

In factories various sanding machines are doing almost the entire work of preparing veneered surfaces for the finishing room, but where the work is to be done by use of hand, scrapers (veneer and cabinet) will give the best results. The skilled cabinet-maker can and will use a plane for smoothing veneered surfaces, but that is more than can be expected of a boy in the school shops. The scraper will be found best for him, even then the work must be done systematically, that is, the work must be started on one end or side, cleaning this off with the necessory number of strokes, proceeding to the other end or side, removing just enough wood to accomplish the object, which is to prepare the surface for sanding. This is then done in the customary manner.

Where water stain is to be used, a sponging down and sanding off of the object, before applying the stain, is worth while. It raises the grain, as well as all compressed places, dented in handling. The raised grain and dents are taken care of in the final sanding. This will leave a surface that is but slightly affected by water or alkaline stain as far as raising the grain is concerned. Little or no sanding of the stained surface is then necessary--this going far to a good finish.

As mahogany is and probably will be the most popular wood for the small objects, that can be made in school shops, a few words regarding the staining of mahogany will perhaps not be out of place.

The author has for years been using potassium

bicromate to obtain the effect of age in this wood. This can be made in any shade desired according to the strength of the solution. Crush the potassium bicromate crystals or salts to a powder and dissolve this in warm water. It is well to experiment on a sample before going ahead with the work of staining because the strength of the solution cannot be judged from the appearance of the stain, which in itself has but a slight tinge of color-- reddish yellow. If the maple, when dry, is found too dark, add more water; if too light, add more of the potassium bicromate. In order to get a definite idea as to how the stain will appear when finished proceed with sample just as you would when finishing the object itself.

The following method has been found to give splendid results: After a liberal coat of stain has been applied and the surplus wiped off, the object is allowed to dry and in doing so the mahogany will turn dark according to the strength of the solution.

When dry, rub in a coat of raw linseed oil, mixed with a little coal oil- -these oils having previously been allowed to extract the coloring matter of alcanet root which gives off a beautiful red color. The oils brighten the wood and brings out the grain. This can then be immediately sanded slightly, if that is found necessary, after which the object is wiped clean and given a coat, of white shellac. Now the final effect of the stain can be seen and appreciated. To complete the finish apply five or six coats of thin white shellac, allowing time for drying between coats. This is finally rubbed down to a smooth finish with a rubbing pad, using powdered pumice stone and rubbing oil.

The success of the above finish on mahogany

is due to the simplicity of application and the fact that the ingredients entering into same have little or no color in themselves.   It really does not matter if the boy makes a few extra strokes with the brush, as this will not affect the finish.

If the object is inlaid with lines or borders of lighter woods, the potassium bicromate stain will not change the color to any extent.   It will give it a yellow tinge, which is not at all objectionable.

FINISH

Milton Keynes UK
Ingram Content Group UK Ltd.
UKHW022033081123
432235UK00005B/120

9 781021 093547